STS-135 Atlantis

Covering the Last Shuttle Ever.

Lloyd Francis Behrendt

Blue Sawtooth Studio

May 27, 2015

Astronaut Fred Gregory, speaking at a ceremony for *Atlantis'* one-year anniversary of its exhibit at Kennedy Space Center, on October 9th, 2014, in response to a question from a journalist:

'everybody gives us the credit...'

Gregory was a three-time shuttle veteran, including flights as Pilot and Commander.

'...and for every launch we had, I think there were probably 40,000 people, around the world, who were going to make it the safest, most successful mission ever.'

Florida Today article by Rick Neale, 10.10.2014

STS-135 *Atlantis* - the Last Shuttle Ever.

Forward

I have been fortunate to have witnessed most of the American Space Program, from its debut launch site at Cape Canaveral, Florida. My family brought me into it when I watched the first rocket ever lofted there, a launch at which my father was the USAF Weather Officer.

Since then I have gone out to the Cape and photographed launches, my library now pushing 400 lift-offs, over 325 of them on B & W film. I shot film until the late 2000's, when I finally shifted to digital.

I am also one of the last artists to paint silver prints with oils.

That unusual mix has led me to this book. In the mid-1990's, after taking a break when our political will failed Apollo, I decided to go back out to the Cape and cover launches just as Matthew Brady photographed the Civil War, with an eye towards History.

I had covered the last four Apollo missions, Apollo 14-17. I left law school to go home after having graduated from the University of Florida, to work as a photojournalist, for my hometown daily paper, the Perry Company's *Melbourne Times*. I left that position shortly after Apollo was cancelled and worked a series of free lance jobs including writing and producing marketing communications, often video documentaries, on the Space Coast.

When I finally got a 'round tuit' and found my way back out to cover launches, the *ISS* construction missions were just beginning. By the time the program ended, I had covered all but one of the 50 or so Shuttle launches it took to lift near 1 million pounds of shelter, gear, and science, 200+ miles up in space.

Though I was terribly disappointed that the STS program was shut down, I promised myself I would indeed 'Matthew Brady' the last mission, STS-135, an unplanned 'add-on' resupply flight, for posterity. *Atlantis* would super-stock the Space Station and use up the remaining STS hardware to mount the mission.

So now I come to this place. It took two years for me to feel it was time to get into the photo lab to make the 20 x 24 silver prints.   Then nearly another two years to paint the images that make up this book. They were selected from the several hundred I shot across the year it took USA (United Space Alliance) to process, mount, launch, recover, and secure the Last Shuttle Ever. I hope you enjoy this look at how we covered it, from the eyes of a rapt observer.

lb,
blue sawtooth studio,
May 27th, 2015

lb photo oil

So it Begins.

Atlantis, having been filled to the brim and then some, was pushed back out of OPF 1. It was just about 8 am Eastern Daylight Time, May 17th, 2011. The Shuttles had been tuned up and stocked for their next missions in the OPF's, the Orbiter Processing Facilities - 1, 2 and 3. The buildings sit just west of the VAB, the Vehicle Assembly Building.

These space shelters are now being re-purposed to support NewSpace: NASA's SLS program capsule - *Orion*; the Air Force' *X-37* mini-shuttle; and Sierra Nevada's *Dream Chaser*, and Boeing's CST-100 for a start.

Atlantis' STS-135 journey began here on Utility Road.

During rollover she made the trip from the processing facility through the Launch Complex 39 Area, to the VAB.

Whenever the Shuttles or any astronauts were around, there was always the silent presence of Security, and their M-16's. A reminder of how the world had devolved and America had become as risky as some third-world countries.

To my knowledge, there was never a proven case of sabotage during the STS program. The extra caution bespoke the gravity of the missions.

Atlantis was designated OV-104, Orbital Vehicle-104. It was by chance, tasked with utilization logistics flight Mission STS-135 (ULF-7), the last Shuttle mission ever:

'Space shuttle Atlantis carried the *Raffaello* multipurpose logistics module to deliver supplies (*lb note*: it stayed on the *ISS*), logistics and spare parts to the International Space Station. The mission also flew a system to investigate the potential for robotically refueling existing spacecraft, and it returned to Earth a failed pump module, to help NASA better understand the failure mechanism, and improve pump designs for future systems.'   *NASA statement*

It was the 33rd flight of *Atlantis*. In its space-faring life, it flew 125,935,769 miles, much of which helped assemble the *ISS*. This is what it looked like as news pros from all over the world showed up to cover the overflowing

lb photo oil

list of activities it took to complete a mission. Several of our Public Affairs Escorts can be seen here in the foreground, watching their charges watch the Shuttle, as it wends its way to space. I firmly believe the Shuttles will come to be judged as some of the most amazing ships ever flown, in spite of the hammering the program came to endure in the media, as politics turned the STS system into a camel drafted by committee. Only then to have a major partner drop out after the Challenger tragedy threatened 'assured access to space'.

In my mind, shutting the program down prematurely threw away at least half the operational life of a 100 billion dollar system. So what if a phone became more capable than the 486 PC's that ran the Shuttles? As space freighters, they worked quite as they were meant to.

lb photo oil

Having watched for years, I felt strongly that shutting down STS was a sin of omission. A well-traveled AF brat hence foreign policy wonk, I worried even before Russia's incursion into Crimea and the Ukraine, about having to be forced to rely on Russian rides for US astronaut's access to space.

Rollover complete, the Shuttle would end up at the north entrance to the VAB, as the Press Corps labored to record the spectacle. This passage offered reporters and photographers the closest approach to an operational bird during actual missions. The shuttles moved at a speed known only to the crews that were their ramrods, according to pre-arranged rules. Sometimes in wait mode, sometimes so fast you missed the shots you needed to get. The Team that 'ran' the shuttles was, in my estimation, one of the best ever assembled. Did anyone else ever build a live-in space station larger than the size of a (US) football field, defying gravity to the tune of (very near) 1,000,000 pounds?

While waiting for my ride back to the Press Site on Landing Day, I got into an air-conditioned pickup at the suggestion of one of the Shuttle workers, to escape the scorching Florida sun. They had seen me sweating my behind off, as one does, mid-day in the summertime. The four guys in the truck all worked on the Shuttle's tiles, and were there to say a final goodbye.

Between them they had over 120 years of experience, ensuring the safety of our men and women heading to space. Amazing. It had become a 'building the Pyramids' kind of project. And all without a ramp to haul the stones.

Here is a shot of me with *Atlantis* at STS-132. It was to have been her last flight, but she got the call for one more. You can see from this picture how close we were allowed to get to the 'Shuttle at Work' during rollover.

lb with *Atlantis* *at STS-132 Rollover*
4.10.2010

photo: Jim Siegel, *Celebration Independent*

Lift to Mate, STS-135 *Atlantis*

Two days later, on May 19th, Atlantis, having made the right hand turn into the VAB, was set for 'lift to mate'. This was the segment of the mission where the Shuttle was attached to the 'Stack' - the mass made up of the External Tank, holding the liquid hydrogen and liquid oxygen that fueled the main engines; and the two SRB's - side-strapped Solid Rocket Boosters.

lb photo oil

I particularly enjoyed this image of the Team using a low tech Hi-Lo to support the early stages of *Atlantis'* last lift. The hoist on the left displays a 175 Ton lift capacity, showing the force necessary to mount Shuttle missions. It took huge efforts to lift that *ISS*-making mass.

The Solids, adding boost, were just what they sounded like - added power to lift the 4.5 million pound launch array. The SRB's were filled with a solid fuel, aluminum, mixed with ammonium perchlorate as oxidizer, giving it an eraser-like consistency.

Unlike the expendable ET, the boosters were built to be re-used. In fact, they are still using them, now to construct and test NASA's next-generation human-rated rocket, the Space Launch System (SLS). Also, unlike the liquid-fueled main engines, SRB's could not be shut down once they were ignited. That was a very big difference. It compounded the risk to crew quite a bit. But SRB's were feasible for re-use, a critical requirement in the original design.

lb photo oil

In order to reach the Stack, the 75 ton Shuttle had to be hoisted up to the top of the VAB by crane. *Atlantis* sat for quite a while on the deck, waiting for things to get lined up for the lift itself. Covering an STS launch was a study in periods of sometimes lengthy waiting, punctuated with rapid, and at launch time, explosive action.

As *Atlantis* finally began to rise, I stepped behind the other photographers to capture the moment. Because I was shooting for posterity by design, I often

lb photo oil

included colleagues, and unfortunately to a lesser degree, the Team, as they went through their paces. Notwithstanding the magnificent hardware, the human talent that worked the missions was the real jewel of the program. I wish I had asked for more access to behind-the-scenes action, but getting badged during the heyday was not certain, especially as a free-lancer.   I kept pretty low key and avoided any such request. In hindsight, it was my bad, but too late now.

lb photo oil

I did not realize the value of the Team until a fellow reporter asked Bill Gerstenmaier, at the time NASA's Manager of the *ISS* Program, at the post-mission briefing: 'After this last Shuttle flight, what have you learned from the STS program?' Gerstenmaier's response floored me. Paraphrased, he said it 'taught him there was nothing we could not do, nowhere we could not go, in space'. Since this Team had overcome the trials and tragedies of building the flying space laboratory that the *ISS* became, there wasn't anything mankind could not achieve, anywhere off of Earth. And the media had lampooned it. Once again in my mind, the conventional wisdom just plain wasn't.

Wise.

Oh well. Just my Early Onset Grumpiness I guess.

Now about mid-way through my time shooting the *ISS* buildout, an injury weakened my leg. Climbing stairs with my gear became out of the question.

So as the lift began, there was the normal pell mell rush of reporters and photographers to get up to the 16th floor, to get pictures from a vantage looking down on *Atlantis*, as she swiftly rose. There are several elevators in the VAB, and you must transfer at certain levels to make your way up higher, the first being at the 5th Floor. Our Escort, having been pressed into duty for the hordes covering the last STS flight, could not find the elevator we needed to get where we were to go. The pool reporters absolutely *needed* to get the shot, so the group quickly 'voted' to climb up 22 flights of stairs, rather than miss the shot. Since I knew I couldn't make that trek up, I had to climb down.

That day I got no shot looking down on *Atlantis* as she went up the last time. Downstairs however, I accidentally ran into Mike Moses, Launch Integration Director, a true polymath, and just plain nice guy. Quite by accident, I got into a one-on-one chat with one of the most brilliant folks I've ever met. From the floor as I waited to meet up with my group of colleagues, I got this image with my digital cam - it was way too dark for available light film.

lb digital image

It became one of my all-time favorite Shuttle shots.

Rollout, STS-135 *Atlantis*

lb photo oil

     Rollout is the part of the process by which the integrated Shuttle and Stack are 'rolled out' to the launch pad, to be put in place for the Big Day, in this case to Pad 39A. Normally the press would follow the Crawler out, but for the last mission we were allowed to observe from upstairs, inside the VAB.

     Here some of my colleagues (the three in the center, l to r, Troy McClellan, Jason Rhian, and Robert Gass) kibitzed as *Atlantis* crawled her way out at 1 1/2 mph, on the same Crawler that had been used to carry out the Moon Program's huge Saturn V's during Apollo. Including the weight of the crawler, the Mobile Launch Platform and the Shuttle, it hauled 12.7 million pounds all up.

     It was now June 1st, five weeks before *Atlantis* began her last journey into space. She was now in place to take off in early July, after all the final preparations were completed.

14

Pad Tour, STS-135 *Atlantis*

About a week later, NASA let us out to crawl all over the gantry and the RSS, the Rotating Service Structure. That day afforded me some of the most dramatic views of *Atlantis* I would shoot. This was taken from the 255 foot level, and shows the port SRB, the ET, and the gangplank leading to the cabin hatch.

lb photo oil

That day, up on L 255, I caught some members of the USA Team working on the ET. This is one of my favorite pics of the entire series, and not just because the guy in the middle is wearing a UF Gator tee. I had captured some of the 'back side' work that had to be carried out in order to make the STS missions happen.

The cone shape, right, is the cap on the port SRB.

lb photo oil

The ET weighed almost 80,000 pounds empty, and 1,670,000 pounds when filled with fuel and oxidizer. It not only provided 'gas' for the Shuttle Main Engines, but it also was the platform that anchored the Orbiter, and attached it to the SRBs, to complete the Stack.

The External Tank was not reusable, and fell back into the Atlantic Ocean, after the Shuttle was set free with explosive charges, once the fuel was depleted.

This is the view from the ground that same morning.

The press corps milling around gave a good idea of the scale of the Shuttle. I owe special thanks to one of my favorite escorts, Pat Christian, center, in the maroon jacket. She was always very helpful to me, including getting me up on top of the VAB to shoot several launches. I will always be grateful to her and to all our PA pals.

This is Pad 39A. It has now been leased to the NewSpace company SpaceX, which should debut its Falcon Heavy expendable rocket off this pad sometime in 2015 if their announced schedule holds. That would be thanks to the vision and aggressive business model that is the brainchild of entrepreneur Elon Musk.   He is an echo of the Robber Barons who punched the Transcontinental Railroad across another new and untamed frontier, the American West. Good for him.

Like the golden stake of yore, Elon's prize is driving humanity away from Earth: his stated goal - the human habitation of Mars.

The next series of images were captured on June 20th, and on July 4th. The crew flew out from Houston in NASA's T-38 chase planes. The former date was the arrival date for the Terminal Countdown Test (TCDT), and the latter, the date just prior to launch. Here you can plainly see another example of the security that deployed each time the crew or the Shuttle was present.

lb photo oil

This was the lead plane at TCDT, tail number 04, piloted by Commander Chris Ferguson, with MS 2, Mission Specialist and Flight Engineer Rex Walheim in the back seat.

The Terminal Countdown Test was a dress rehearsal for the actual launch. The astronauts would suit up, drive out to the pad, get strapped in, and take the countdown all the way to T-0, so they would be familiar with the exact process when it came to the real thing on Launch day.

This image was shot over the shoulder of NASA videographer Ben Smegelsky as he was capturing the arrival of Chris Ferguson and MS 1, Mission Specialist 1 Sandy Magnus, in T-38 tail number 61.

It was the 4th of July, 2011.

It was just four days before launch Launch Day, July 8th, with a T-0 set for 11:26 EDT in the morning.

The photo opportunity at the landing strip was about the only guaranteed chance for the Press to get up close and personal with the astronauts just prior to launch. The Crew were in quarantine at this point, and no chances were being taken, that one might catch a bug.

The T-38's herald the approaching launch, and showcased the skills of the STS crew leaders. The flying involved in these trips was another means of keeping flight skills sharp. And it was effective shuttling astronauts back and forth from Mission Control in Houston to the launch site at Kennedy.

On this last mission, the Pilot was Doug Hurley , the Commander, Chris Ferguson.

lb photo oil

Shuttle Launch Director Mike Leinbach welcomed Chris and Sandy as they deplaned their T-38, and the NASA cameras rolled to save the action for the world to see.

Shuttle crews then traipsed over to the waiting press, photogs and reporters assembled in a line, usually in the Florida heat, separated from the astronauts by a rope barrier.

It was a chance for Crew to make brief comments about the upcoming mission to inform the public.

We always knew what was up, as we had the passion and responsibility to know what the mission was about.

This setting for us then, would provide sound bites and photo ops that we could share for those not fortunate enough to be there.

I always considered it a huge honor.

Doug Hurley, left, was a Marine Aviator and F-18 Hornet Test Pilot, before his selection by NASA in 2000.

lb photo oil

Our paths had crossed by inches when he was the Lead ASP - Astronaut Support Personnel - for STS-107 Columbia. He had quickly shepherded the 107 crew's families into waiting cars just as I stumbled by them, a confused reporter, rushing to catch my bus back to the Press Site. That day, the Shuttle had failed to return.

Hurley would cross paths with History again. He was selected as the last Shuttle Pilot - on *Atlantis* STS-135.

Chris Ferguson, on the right, was a Navy Captain and Test pilot who flew F-14 Tomcats. He was Pilot on *Atlanti*s for his first mission, STS-115, and Commander on its last, STS-135. He now is now Vice President for Boeing's CCP - Commercial Crew Program - Division.

Sandy Magnus, speaking here, was a McDonnell Douglas engineer who worked on stealth aircraft, and later was selected as a Mission Specialist for STS-135, the standby rescue mission for STS-134.

Successful completion of STS-134 paved the way for her to fly on the now-scheduled last Shuttle mission.

Rex Walheim, to the right of Sandy, helped shake down some systems on the Shuttle for the USAF, and then went on to become an instructor at the USAF's Test Pilot School, before being selected as a NASA astronaut. He had over 35 hours spacewalking outside the *ISS*, a function of his Mission Specialist duties on several Shuttle missions.

He was Flight Engineer and MS 2 to Sandy's MS 1.

This final T-38 flight out of Houston for the astronauts prior to launch was Independence Day, July 4th 2011. Note the flags, if you will.

T-4 days until launch.

lb photo oil

the 4th of July, 2011

After brief remarks, the crew went to the Ops and Checkout building to mark time til launch. We were once again back at the Press Site, one last chance to tell the World about the end days of the Space Transportation System.

The Shuttle.

Atlantis OV-104, weighed 151,315 coming off the line in Palmdale. It was lighter than Columbia by 3 1/2 tons.

She was named after another workhorse, a two-masted 460 ton ketch that was operated by Woods Hole Oceanographic Institute from 1930 to 1966. That wooden ship was the first U.S. vessel to use electronics to map the ocean floor. In the annals of human in space, the Shuttle was King of the Hill in terms of the amount of payload it lofted up to weightlessness, 200+ miles up. Atlantis lifted about 30,000 pounds on her last flight, and fully 20,000 of that was payload. It orbited the Earth 200 times, lifting the last American-soil delivered care package for the next four years.

lb photo oil

This shows *Atlantis* on the pad during RSS retract, where they roll the Rotating Service Structure off to the side, preparations now mostly complete. It opens the Shuttle to the pad and makes for good photos. The crushed rock in the foreground is the crawler-way that the Shuttle rolled in on, and the Gate to the right was the guardian of the rocket ship. The system was fragile in some respects - no FOD, Foreign Object Debris allowed: we would have to 'croakie' strap our sunglasses on so they wouldn't fall off as we clambered around the pad. No chance could be taken on a lost pair bouncing around and chipping a tile at launch time.

Yet in other respects, Shuttle was as robust as any barge ever made. It lifted nearly 1,000,000 pounds of payload total, orbiting the earth at an average altitude of 220 miles. It weighed almost 5,000,000 pounds each time it lifted off. It was a machine of immense proportion. And as such, it made possible amazing science, amazing discoveries, and amazing knowledge about long stays in space.

Quid Pro Quo. But for the Shuttle, there could be no *ISS*.

Very hard to dispute that.

This was a team of students and their Principal Investigator, Cheryl Nickerson, from Arizona State University. They were celebrating the impending launch of their experiment, continuing medical research on the *ISS*.

At the time, their work was hitting nearer and nearer the bullseye for a vaccine intended to inoculate against an organism that sickens and kills many people each year - the salmonella bacteria, the culprit in food poisoning.

It was now L-2 Day, July 6th, two days before launch.

Things would be rocking and rolling at the Press Site. There was always excitement, and swag, and bs-ing with your pals.

So back on the bus, Cadets! The PA folk were herding cats on this one and they knew it.

Still our NASA escorts made it happen for us.

lb photo oil

After a kind of cloudy trip out to the pad for RSS Retract, it was back to the Press Site for press conferences. The STS-135 pre-launch briefing starred a tight group of mission managers:

Mike Moses, speaking here, was the Launch Integration Director; accompanied by Mike Leinbach, Shuttle Launch Director, center; and Kathy Winters, Shuttle Launch Weather Officer.

Mike Moses had all the attributes of an Expert with a capital E. If you heard him talk about the Shuttle, it seemed there wasn't anything he didn't know. His report? We were on target for L Day. So far.

Mike Leinbach had an amazing knowledge base as well, and was as calm as he was adept at proctoring the Shuttle through its mission paces.   He echoed Mose's assessment and then took some questions. Looking good so far...

Last up, the Weatherman, er Weatherwoman, er Shuttle LaunchWeather Officer Katie Winters: not so optimistic as were the previous two presenters.

Due to the potential for showers and isolated thunderstorms in the area, there was only a 40% chance of acceptable weather at launch time. It *was* summertime in Florida, and daily showers were likely that time of year.

The Shuttle had a launch constraint that other rockets did not suffer. Other rockets normally launch if there is no lightning potential within five nautical miles, to keep from possibly blowing up. The Shuttle had to always anticipate a RTLS maneuver - Return To Launch Site. That meant two things. First, a very heavy glider with *no power,* it had only one shot at landing. And second, because of that, it had to have line-of-sight with the runway at all times. It needed a 20 mile radius, free from *any* clouds obscuring the SLF (Shuttle Landing Facility).

But 40/60 wasn't awful, especially for summer time.

It would be the luck of the draw if she went on time on L Day.

L-2 Day: Press Site 'Carnie' Day, STS-135 *Atlantis*

Most often the Press Site at the Kennedy Space Center was pretty quiet, just a few full rows in the big parking lot would be filled with the cars of the regular NASA Public Affairs employees and beat reporters who called the place home as they carried out their important tasks. It then fell to NASA PA to make sure whatever visiting press asked for at mission time was accommodated, ensuring as much as possible that the word got out about the latest stuff going on at KSC. And of course, every few months, coping with the high energy nature at launch time itself.

That's when in came the satellite trucks, the foreign reporters, and the school kids, along with your 'regular' colleagues from the last one, and those few who 'knew somebody'.

For this last time, there were people everywhere, setting up like the Carnival had just hit town, preparing to thrill the crowds that would appear in time for the fireworks.

They streamed into town filling up the nearest hotels, and after setting up, they headed for their favorite watering hole to wait some, and see who knew the most about the latest space news.

Here is a glimpse of the eclectic chaos that emerged at this point, the excitement and tension of the pre-launch wait creating a constant buzz.

lb photo oil

At the top of the hill stood this crew setting up for NASA's Digital Learning Network. Looks like there were two grips, the audio guy, the camera man, the remote producer, the director, the lighting guy, and the astronaut star (who I subsequently found out while researching this image) was Leland Melvin, center.

Elmo must have been taking a nap. He was the real star. Included in his NASA clip was the STS-135 launch, which you can see at this link.

https://www.youtube.com/watch?v=zofSItNeAlI

I'm figuring it was shot by this crew. It's called 'Learning Space: Elmo Visits Kennedy Space Center'.

Running time is 1:33.

That is what it looked like from the top of the hill gazing at the Shuttle on Pad 39A with very fancy cameras. This photog was pre-setting her shot, surely not leaving this important assignment to chance.   As my Dad 'the Col.' liked to say, 'Prior Planning Prevents Poor Performance - the 5 'P's'.

lb photo oil

The air was electric, the wind-up palpable and building all the time.

More folks were arriving, more tents going up to cut the summer heat, more bare tripods set up, each staking out and holding a photographer's spot.

My spot was always on the northwest corner of the small concrete pad that supported the flagpole by the Countdown Clock. I had kind of claimed that spot back in the mid-120's of the STS program. I had decided to move up next to Carleton Bailie, as he was most always set up just southeast of the flagpole.

*** *** ***

You checked your exposure and focus one more time.

So set up and ready, we were.

All there was left to do was to light that sucker.

Launch Day, STS-135, 8 am Eastern Daylight Time

It was early on Monday morning, July 8th, 2011. The field on the east side of the press site was almost as full as when huge crowds used to show up for Apollo launches.

For us here today, one last shot at glory.

I had been tapped to be the NPR 'Minute of the Day' on their morning show airing on Launch Day. Someone had heard me yakking quite by accident on a prior CNN podcast for STS-134, that I figured was pretty insignificant. but it got me the minute for this day.

In retrospect, for the final minute the editors took out of our 20-something minute interview, I thought I kind of sounded full of myself. During our visit they steered me towards saying I would cry, that was not my first response. The media seemed intent on trying to portray the whole program as gone, when the fact was that the STS program was a large, but *partial*, contributor to Cape launches. Good stories don't sell papers I guess. I had been told four days before, by the 45th Space Wing Commander, General Ev Thomas, that *five of the six working pads had active missions* on them the day after the launch of STS-135..

Launch Day, STS-135, 11:25 am Eastern Daylight Time

You could cut the anticipation with a knife. The air was as thick as the humidity with it. Shuttle launches were always that way,. The more people, the more you sensed that tension, like a bow's string pulled all the way taut, ready to snap.

The heat had already set in but you paid it no mind. You knew the upcoming show was the ultimate distraction.

The Lucky Hat was in place on my head, me smack right up next to the flagpole, at our regular spot...

The clock was counting down:

George Diller, the voice of NASA for STS at KSC, was expertly describing the pre-launch events as we got closer to liftoff:

'T-1 minute and counting...

T-40 seconds, handing off to *Atlantis*'s computers at T-31...

Female Voice: 'Call for a hold at T-31 seconds due to a failure...'

Your first thought is always, oh no, what now, even though you know it is rocket science. Lives are at stake - you use full caution.

And if you are out at the Cape all the time, you know things can happen.   Now with almost-launch time tension racheting up second by second, something unexpected this late in the count was messing with the flow of things.

As it turned out, after a brief hold, things were back on track. Backup cams needed to visually verify a 'beanie' cap that secured the ET's tip had moved out of the way. It had, and within 3 minutes, George had begun again:

'handoff to *Atlantis*'s computers has occurred...
Firing Chain is armed...
Go for Main Engine start...
T-10... 9... 8... 7... 6... 5...
we have Main Engine start. All three engines up and burning,
2...1... ZERO!...
...and Liftoff! the final mission of *Atlantis*, on the shoulders of the Space Shuttle, America will continue the dream!'

for the last time, George Diller - STS-135 liftoff voice-over.

the Launch of *Atlantis*, STS-135

lb photo oil

31

You can watch the launch here, courtesy of the cloud.

https://www.youtube.com/watch?v=VjeUckqWxgo

As for me, I was as elated as always. There were no tears, PBS, when *Atlantis* took off at 11:29:04. Only three minutes or so after the planned 11:26:06 liftoff time, close enough for the Lucky Hat to take credit! The hold had been very short, allowing the nearly on-time launch.

I hooted and hollered and pumped my fist in the air. The huge field east of the press site, with its flag and countdown clock, was slammed chock full with amazingly fortunate people. It was the crowd of awestruck folks who made it there that day, lucky enough to wheedle their way out onto the Cape for the last, glorious, historic, immensely physical, crackling, launch of *Atlantis*.

And by default, the last of the Space Transportation System.

Flight Day Four, STS-135 *Atlantis*

lb photo oil

You can tell how still it was by the fog that had settled behind the *Liberty Star* as she made her way back from the recovery zone, 140 miles offshore.

She carried a crew of about 20, if you included the eight divers that were on board. It was their responsibility to secure the booster by first hauling in the chutes, and then diving down the SRB about a hundred and fifty feet, and capping the business end with a 1400 pound plug. That allowed them to pump in air, floating the behemoth to a horizontal position, one that *Liberty Star* could then tow back to the Cape.

As I shot the returning SRB, I was about to realize this mission had become a true right of passage for me as well. I discovered an old back complaint this day had turned into the inability to walk normally.

Like the Shuttles, my age was beginning to impact my utility.

I found myself out on the tip of a jetty that juts 1200 feet into the Atlantic Ocean, shooting with a gaggle of career-long colleagues, some since the 70's. Now that the SRB had passed the first point that images could be captured, that herd was off in a cloud of dust, to follow it through the locks. I quickly discovered there was no way I could keep up.

My saving grace *that* day was based on the uneven flow of the return process.

NASA decided to stay on the sea side of the locks for the night, as they shifted the SRB. Using my journalist radar, I had a good idea where I might find my shooter buddies, and I managed to catch up with them at Fish Lips, where we all filled up with some nutritious substances. Then all of a sudden, NASA, in their own inimitable way, decided they were going to move the SRB through the locks - right then.

photo Dawn Leek Taylor

Just after I had consumed two of these in order to properly re-hydrate.

It would have been criminal to waste any.

Quick-buzz, combined with getting beat out in the 'hunt' by some lookielous who were quicker than me in the chase, made me drive like a madman in a NASCAR race, thru the narrow, rutted, winding two-lane road that takes you to the locks. Much too fast. I made it without event. Thank you Jesus.

The *Liberty Star* got to the locks just as I pulled up. I left my Jeep Cherokee (also an *ISS* STS veteran) parked in the middle of the road, the Guinness surely having affected my judgement.

photo Mark Usciak

It looks like I was moving out in this pic, but I was still late to the dance, and missed getting a decent shot of the *Liberty Star* just as it passed through the Locks. I had not the attention span to find all my colleagues in the crowd there, but I am pretty sure these folks were: Mark Usciak, Julian Leek and Dawn Leek Taylor, Matt Travis, Mike Barrett, and Walt Scriptunas II - the owners of those cams back there on the table at Fish Lips. It was in the very early days of Facebook, and someone contacted me there afterwards, saying they saw my beer-fueled lame-butt race to the Locks and stated 'so that's how a pro does it.' I cringed and told him no, very poor judgment. Never again, lb.

Locks finally underfoot, I had a great spot from which to visually explore the amazing geometry the Piers offer up.

lb photo oil

This is the 1400 pound plug that I spoke of earlier.

This image series will give you an idea of the length of the SRB if you compare it to the size of the boat towing it, the *Motor Vessel Liberty Star.*

Her measurements: 176 feet long, 37 feet wide, drawing 12 feet of water.

She had two conventional engines and backup movable jet propulsion, the latter used in order to protect both endangered manatees in the lagoon, and the divers out in the Atlantic, during booster recovery.

I thought that was pretty cool.

Maybe they are the same type thrusters the SpaceX Drone landing pad barge uses as they perfect landing the first stages from Falcon 9 launches?

Shades of STS. Parallel creativity.

Legacy.

lb photo oil

It had been ages since I was up at the locks, having come up once or twice to try and get pictures of some manatees there, without much success.

The scenic, however, was an artist's dream, a palate of perspectives and reflections.

It was surreal to think I was playing witness to one of the most magnificent engineering projects in history, at its end of days.

And of course, once again, the press corps was reduced to the handful of regulars, who faithfully covered *all parts* of those other STS launches.

We were predictably there, to capture the less sexy moments as well, away from the flash and thunder of Launch Day.

For me these turned out to be some of the most fun to work with on the bench.

I wonder what share of attention they will garner in the exhibit that is this book.

Art imitating life?

Landing, STS-135 *Atlantis*

Several things happened that made Landing Day, Mission Day 14, not my most fun ever. It began when NASA managers chose not to push one orbit so *Atlantis* could land just after sunrise, instead of in the dark. I am sure in the annals of space travel, one does not change plans for such a seemingly superfluous reason. Indeed most of the digital photogs got decent shots, like this one by Mark Usciak. Due to the overflowing last-mission crowd, Mark stood for his shot practically underneath the legs of my tripod - his image was, per his usual professionalism, superlative.

photo: Mark Usciak

Atlantis, the last Shuttle ever, had touched down at 5:57:00 am EDT, July 21st, 2011.

I had shot enough landings to know where the primo shot was, a 'hole' as it were, in the invasive bushes that had grown up to obscure the landing strip. Mark's shot is very close to what my shot would look like except for two things.

First, since I used B & W film back then, it was too dark to get much of an image in the pitch black. I had never come out to even try a night landing, so I was doing one in the dark for the first time. No experience loop to guide me.

Consequently, I was completely untested in that set of circumstances.

Secondly, had I really thought about it, I would have known to start shooting before the Shuttle reached the sweet spot in my viewfinder. It took me two years to figure it out: that due to my normal reaction time, about 1/8 of a second, I totally missed getting the bird in the frame, and got just bare runway instead. By the time I had seen it in the viewfinder and clicked, it had sped by at landing speed, over 200 mph.

Gone, gone, gone.

No do-overs on that one, no landing shot for lb.

Dang.

My second misfortune on Landing Day was as follows:

From: "(KSC-PA000)"
Subject: STS-135 Landing Runway Photo Opportunity Status
Date: July 21, 2011 12:07:18 AM EDT
Cc: (KSC-PA000)"

Due to the limited number of available slots, we regret we are unable to accommodate your request to participate in the runway photo opportunity after shuttle Atlantis' landing. Video b-roll and images will be available. We appreciate your understanding.

After all the STS missions I had covered, over 50, my crew and I were culled *out* of the group of journalists (and some not-so-journalists) who were selected to go out to the runway to be with *Atlantis* and the STS-135 Crew on the runway, as they came home for the very last time. That photo op had always been strictly reserved for the pool photographers - wire service journalists - before.

To this day, my colleagues tease me about "no 'red dot', lb" - the pass one needed to go out to the runway post-landing.

I remember the stunned feeling upon reading that email, at 7 am after being up all night waiting on landing. There were folks who went out who were undeserving, and we had paid our dues. At least that's the way it seemed to me.

I do remember thinking it was a real slap in the face to be passed over.

I never made a fuss out there, and I likely would do nothing different today.

As a result however, History robbed me, that day, of film images of that unique occurrence.

Towback, STS-135 *Atlantis*

Towback was something a photographer looked forward to. If the wind was slack, there was a nice reflection shot from the canal in the foreground, with the parade of landing keepers strung out in the background. It also is a fine illustration about how the Shuttle got back to the hangar from the air strip.

lb photo oil

The Shuttles were scarred, but magnificent ships to be around. It was hard not to be in awe when we got to get close enough to touch, even though we knew better. *Columbia, Challenger, Endeavour, Discovery* and *Atlantis* - they all made this trek when they returned to the Kennedy Space Center - whether they landed here or were ferried in on NASA's 747 Shuttle Carrier Aircraft.

It always seemed to provide a fitting end, one that gave the birds a chance to show off a bit for the crew that had launched them.

In this case, 14 days before and 5,284,862 miles ago.

lb photo oil

So the Space Shuttle *Atlantis* had capped off the low earth orbit tug business, forever.

Or at least for a long, long, while.

There will likely never again be the will to do such an amazing thing.

We now have space taxis of some sort, but chances are very slim we will ever again see a flying ship of this size, with which to build miracles up in space.

This shot for me, gathers the pride of the STS mission all in one place: the USA employee (and his Lucky Hat) in the foreground at right, another team member carrying a couple of waters out to his ground crew pals, and of course, the pretty lady *Atlantis*, herself.

It was about 11 am EDT, July 21st, 2011.

Final Wheelstop at the OPF.

Press Coverage, STS-135 *Atlantis*

You are seeing the working area of the Kennedy Space Center Press Site from a vantage I enjoyed across the last 50 Shuttle missions.

I made it my good fortune to finally get back out and cover launches at the Cape, after a shooting hiatus that was triggered by the demise of the Apollo program.

lb photo oil

I had finally gone back out for the Mars Pathfinder, Delta 2 launch, in November of 1996.

Why? Because America was finally *going* somewhere again - a mission to Mars. Pathfinder went on to be the first flash crowd on the internet, when everyone wanted to watch the Mars landscape go by live, as Pathfinder putted around on the Red Planet.

And that crashed NASA's servers, another net first.

lb photo oil

That return to shooting the 'Air Force side' launches in 1996 led 'photo-me' to wander across the NASA Causeway to the Kennedy Space Center just as they were flying the Shuttle to MIR in preparation for the building of the International Space Station. And then I continued watching in awe, all the way to the completion of an amazing new Pyramid - built way up in the heavens.

It was the chance of a lifetime. And I was going to do it justice.

Along the STS trek, I would drive up I-95 every three or four months, at all times of the day and night, to watch astoundingly magical launches of one of the biggest rockets ever built, from a mere 3 miles away.

It was always interesting, seeing folks you knew in the trenches every so often, kind of like family reunions.

I am usually very good with names but there were so many whose faces I knew so well, we would nod hello each time, but their names would escape me. I was usually too embarrassed to ask about something courtesy should already have secured.

But there was also a tight circle of usual suspects.

lb photo oil

The waiting, the asking, the reporting - the Press Site is where that all happened. It was an open book, by order of the People of the United States, to reporters, photographers, and space wonks who had figured a way to get out to the Cape.   They made themselves a serial workplace there, for the duration of the launch. Sometimes it was for just a few hours. Sometimes it spanned days, and sometimes even months. If it was long enough, the group assembled, would retreat to their hotel rooms or their regular homes, to once again hurry up and wait for the techs and engineers to get it 'fixed for launch'.

This is where we hung out, good-naturedly trying to out-maneuver each other in a sporting competition to see who had the latest and most accurate insider's tidbit: why the launch was delayed, what would fix it, and of course, how that affected when the launch would actually go.

It was usually Ben Cooper with the scoop. He had come from NYC and then schooled at Embry Riddle. Very inside information from Ben.

He is now shooting for SpaceX I think, so there you go!

He, and Justin Ray of SpaceflightNow.com, and Stephen Clark, also at SFN, seemed to know what was up before some of the folks who should have.

That kind of information was physics too. Flow was flow, always finding a path of least resistance. Information had its own black market.

And its consequences.

It was an unwritten rule you didn't make any pronouncement before it was announced by an official. As it should be. Might impact future credentials.

It was in that big room, with rows of reporters benches, that we were able to ask questions of experts from all the far-flung places, flown in to support their piece of the mission. Often it gave us the astronauts themselves.

It was the place where I spent many nights nodding off, pre-launch. And it was also where I stood in the back corner, taking pictures of all the rest of the Press Corps, watching *Columbia* STS-107 stream down in pieces, terribly, across Texas skies. They were all huddled around the lone 18" TV up front, the only one in the room at the time. The History of that instant was gripping.

I still go up there for NASA's current events, leading up to future NASA manned launches, and also for the EELV (Evolved Expendable Launch Vehicle) science launches they mount from time to time on an Atlas V, or a Delta 4, or a Falcon 9. I hope I am there in a few years when America again launches our astronauts into space on our rockets. Maybe we can get it right this time.

They say third time is the charm.

the End, STS-135 *Atlantis*

The German heritage in me hates to admit failure if an effort comes up wanting. Or is that just human nature in general?

lb photo oil

Hence my non-picture of the landing here.

One nice thing about photo oils is that you can sketch a bit if it helps, though I do so quite sparingly.

This was an instance where that would be the only way for me to cobble a meaning that would work, a way to snatch victory from defeat, in my battle to tell the News.

You cannot tell from my non-landing image, but I can: it is the exact same spot seen in the landing picture Mark Usciak got, displayed here a few pages back. It's only that mine came about an 1/8th of a second later...

In that span of time, for my stead, the STS program went from 'there', to 'not there'. It had become an analog equivalent of a digital 1 and a digital 0.

The bucket that had seemed so full was all of a sudden so very empty.

I will always have opinions about those who squandered the great momentum we had gained, in our fitful attempts to tame space across the years.

Who's to blame doesn't change the fact, that in the arena of American exploration, the politics of the achievement did not keep up with the wishes of the American people.

The program had ended with the same capability for carrying Americans to space as the hollow echo my landing picture (an error) mirrored. That is to say:

None.

Still, NASA has saved the SRB segments, and Shuttle Main Engines, for use with future designs. And I was recently reading a NASA release about heat tiles being attached to Boeing's *Orion* capsule, for its now-successful 2014 re-entry test flight (unmanned of course) on an Delta 4 Heavy. Perhaps it will ride on the SLS - the Space Launch System - an Apollo-like vehicle, heavy lift, for long-distance manned space exploration.

And that can work fine.

Except for their shape, the tiles look just like the heat tiles that were perfected during the Shuttle program. So pieces, parts, and technologies drift forward through the meandering program, being recycled as we again try to find a good way to reach our best future.

And wait to see which way the wind will blow.

Not surprisingly, any political will still seems fragmented. In my humble opinion, it has been a heritage sadly squandered.

So far.

I hold out a hope however, that a champion, a leader with the Right Stuff, will emerge to carry the banner of American know-how and drive.

There are plenty smart and brave folks out there who could.

In the meantime, we will be here scribing the transition. We are go...

See ya' at the next one!

Lloyd Francis Behrendt
blue sawtooth studio
Malabar, Florida

epilogue

I have just finished another edit and graphic punch up, a last pre-publishing task that was interrupted by a healthy dose of back surgery.

That has prevented me from shooting any launches at the Cape and will for a while, thought I have just cleared myself to go out in July 2015. Of course it is the busiest it has been in years with at least 24 launches scheduled in 2015. But I am counting on getting back out to continue my never-ending hunt to document our space program.

Let's see what else has happened?

Elon and SpaceX, having now launched 15 Falcon 9's here. They just had a failure of a Falcon 9 (NASA CRS 7) but they will be back in the saddle quickly I predict. They are well along the way to modifying Pad 39A, where STS-135 launched, to handle the Falcon Heavy, made up of three Falcon 9 cores strapped together. It will be the most powerful launch system in the world. They announce that they are hoping to launch one before 2015 ends. They just announced they will do their second launch abort test from 39A rather than Vandenberg. Think they found out how real the Cape is when the chips are down.

And having tried 3 times offshore, SpaceX now has their own assigned Landing Pad, Landing Pad 1. I think that is the coolest thing! They are practicing landing offshore on a re-purposed oil platform, that is ported at Jacksonville, the first named 'Just Read the Instructions', the new one, 'Of Course I Still Love You. Gotta like their style!

Just this week, the local economic development group announced Blue Origin, Jeff Bezos' (Amazon) effort, is looking to site manufacturing operations at the entrance to KSC, and would use Pad 36 as a launch site.

So the bs that was the Space Kool Aid of 2011, that the Cape was 'done after STS', never came to pass. Boo-Yah!

epilogue 2

Two days later and the County Commissioners just granted $8 million in tax abatements that will locate 900 aerospace jobs here in launch/manufacturing: 600 for Blue Origin, 300 for Lockheed (Trident Submarine Operations).

epilogue 3

'We Are Go...' new promo campaign out of Space Florida touting the Cape for space tourism. Just met Eric Garvey, the new head of the Tourist Development Council. He is so hip on space he is (thank you finally!) organizing launch parties as area events. Takes someone from outside to see that.

epilogue 4 ... 3 months out from back surgery, just decided I am walking (the right way again!) well enough to go out in July to cover the thee ULA launches that are scehduled, two Atlas Vs and a Delta 4.

et cetera

Escorts and PA folks, in no particular order: Gail 'Stormy' Villanueva Bell, Bill Wilson, Manny Virata, Pat Polidora-Christian, Laurel Lichtenberger, Johnny Johnson, Norris Gray, Jennifer Horner, Kathleen Ellis, Ken Thornton, Susan Wells, Jessica Rye, Major Greg Harland, Capt. Warren Comer, Capt. Kevin Coffman, Dr. Sonny Witt, Christopher Calkins, Col. Mike Rein, Lt. Col. Pat Barrett, Maggie Gwinner Persinger, William Johnson, Julie Andrews, Hugh Harris, Sarah McNulty, Mike Curie, Trina Helquist, Jack King, Charlie Parker, Mitch Varnes, and ...

Colleagues, inpo: Carleton Bailie, Craig Bailey, Mark Usciak, Alan Walters, Julian Leek, Dawn LeekTaylor, Christopher Taylor, Keith Rudroff, John Connors, Terry Reilly, Mike Killian, Ben Cooper, Stephen Clark, Stephen Young, Justin Ray, Matt Travis, Tom Usciak, Roger Scruggs, Craig Rubadoux, Tim Shortt, Malcolm Denemark, Scott Maclay, John Studwell, Mike Barrett, Stefano Coledan, Pat Duggins, Bob Gass, Troy McClellan, Charles Twine, Vicki Barnes, Joe Marino, Bill Cantrell, Bill Harwood, Dan Billow, James N.Brown, GBI, Walt Scriptunas II, Sandy Frederick, Mike Howard, Melanie Lee, Billy Cox, Barbara Zelon, Curtis Krueger, Suresh Atapattu, Joe Sekora, Beth Dickey, Bill Ingalls, Ben Smegelsky, Jason Rhian, Amy Melissa O'Brien, Linnea Edmeier, Gene Blevins, Bill Hartenstein, Zuzanna Falzmann, Jim Banke, Robert Pearlman, Ken Havekotte, Peter King, Matt Stroshane, Ken Strohm, George Bell, Jim Siegel, Tom Rogers, Todd Halvorson, Emily Carney, Graham Martin, Gerhard Daum, Val Phillips, Sam Gordon, Jeff Siebert, and ...

All research done at NASA or Wikipedia, on the net.
Copy Editing, Susan Harrison
Technical Editing, Mark Kirkman
Additional photography: Mark Usciak, Dawn Leek Taylor, Terry Reilly.

blue sawtooth studio
1085 Hall Road, Malabar Florida USA, 32950
lbehrendt@cfl.rr.com
www.facebook.com/SpaceBrat1
copyright blue sawtooth studio 2015

www.ingramcontent.com/pod-product-compliance
Lightning Source LLC
Chambersburg PA
CBHW050902180526
45159CB00007B/2764